KEVIN MAGNUSSEN BIOGRAPHY

The Viking Warrior of Formula 1

Fred K. Edwards

Kevin Magnussen Biography

Copyright ©Fred K. Edwards. 2024

All rights reserved. No part of this publication may be reproduced, distributed, or transmitted in any form or by any means, including photocopying, recording, or other electronic or mechanical methods, without the prior written permission of the publisher, except in the case of brief quotations embodied in critical reviews and certain other noncommercial uses permitted by copyright law.

Kevin Magnussen Biography

TABLE OF CONTENT

Kevin Magnussen Biography

INTRODUCTION

Few drivers in the high-octane realm of Formula 1 truly represent the tenacity. They will need more in this sport than Kevin Magnussen. Known as the "Viking Warrior," Magnussen's ascent through the motorsport ranks is a story of tenacity, skill, and a relentless pursuit of excellence. From his early days karting in Denmark to negotiating the challenging circuits of the F1 season, this biography traces the life of a driver who has often overcome the odds.

Kevin's narrative is not only about racing; it's a monument to passion, family legacy, and the unrelenting search for one's aspirations with a fierce competitive edge and the heart of a fighter. Readers will explore the difficulties he encountered, the victories he enjoyed, and the steadfast Viking spirit that propels him ahead as he battles wheel-to-wheel with the greatest of the world. This biography is an inspirational narrative of a modern-day warrior whose legacy in Formula 1 is still

Kevin Magnussen Biography

developing, not only a recounting of races. Discover with us Kevin Magnussen's life and career a real champion both on and off the track.

CHAPTER 1: WHO IS KEVIN MAGNUSSEN?

Danish racing driver Kevin Magnussen is well-known for his determination and Formula 1-oriented skillfulness. Born on October 5, 1992, in Roskilde, Denmark, he hails from a family steeped in motorsports; his father, Jan Magnussen, is a former F1 driver and accomplished competitor in several racing series.

Magnussen started his racing career in karting, where he rapidly proved his ability. Rising through the ranks, he won multiple championships, including the elite Danish Formula Ford series. He relocated to the UK to participate in the British Formula Renault series in 2010, where he won the title in 2011, therefore displaying his future motorsport ability.

Kevin started Formula 1 with the McLaren team in 2014. Having a podium finish in the Australian Grand Prix, his

Kevin Magnussen Biography

debut season was outstanding, and he became the first Danish driver to stand on the podium in F1 since his father. Magnussen developed a reputation as a fierce competitor from his aggressive driving approach and outstanding accomplishments.

Following his time with McLaren, he joined the Renault team, where he continued to display his abilities. He has driven for numerous teams throughout the years, including the Haas F1 Team, therefore proving his relevance in the sport. Magnussen, well-known for his tenacious work ethic, willfulness, and ability to squeeze performance from his car, has developed a reputation as a powerful driver on the grid.

Kevin is well-known, of course, for his lively demeanor and close relationship with his supporters. Still dedicated to the sport, he constantly aims for excellence and welcomes the difficulties of driving Formula 1. He is a well-known person in modern motorsport since his path embodies not just the search for speed but also the warrior attitude.

Kevin Magnussen Biography

Childhood in Denmark and early motorsport love

Kevin Magnussen's early years in Denmark were rich in motorsport, mostly shaped by his family's background. Denmark is also home to Kevin, who was early exposed to racing after being born to famed former Formula 1 driver Jan Magnussen and his mother, a pillar of support throughout his life. Growing up in Roskilde, he was surrounded by tales of competitiveness and speed that stoked his love of motorsports.

Kevin seemed drawn to everything with wheels from the time he could walk. At eight, he first had major exposure to racing through karting. Often, his father would take him to races where the young Magnussen would marvel at drivers negotiating tight turns and accelerating down straightaways. Kevin started riding a go-kart not too long ago. With his extraordinary skill and fierce competitiveness, he soon proved to be a natural and attracted the interest of everyone around.

Kevin Magnussen Biography

Early Kevin's karting years were full of contests both inside Denmark and abroad. His commitment and diligence paid off when he started to accumulate wins in national and local titles. Industry insiders noticed his outstanding accomplishments, which created possibilities to race in additional elite tournaments.

The Magnussen family gave up things to support Kevin's growing profession. For contests, they flew throughout Europe, devoting time and money to helping him as a driver. His strong work ethic, resilience, and the value of teamwork qualities his parents taught him would help him all through his racing career.

By the time he turned fifteen, Kevin had already established a strong reputation in the karting scene. He then switched to motor racing, running in the Danish Formula Ford series, where he soon became a fierce contender. Early success in these formative years not only stoked his love of motorsports but also set the path for his career in Formula 1.

Kevin Magnussen Biography

Kevin Magnussen's early years were more than just racing; they were a mix of family support, an unrelenting quest for excellence, and the excitement of rivalry. This basis helped him to become the Viking Warrior he is now, prepared to face the demands of the greatest level of motorsport.

Karting Beginning, And The First Taste Of Competition

Kevin Magnussen's karting roots were crucial in determining his professional racing driving future. At the tender age of eight, he began karting and fell fast in love with the sport. Spending the first few years developing his abilities on several circuits throughout Denmark, he would regularly challenge more seasoned and older drivers.

Early in Kevin's career, his father, former Formula 1 driver Jan Magnussen, was absolutely vital. Jan shared insights from his own racing experiences, thereby acting

Kevin Magnussen Biography

as a mentor and offering direction. Kevin could thrive in this environment fostered by family support, which nourished his passion and will.

Kevin started competing in national and international karting events as he moved through the rankings. His first major taste of rivalry came in the Danish Karting Championship, where he soon established himself. Often finishing on the podium and winning the respect of his colleagues, Kevin displayed amazing speed and ability.

2009 was one of the turning points in his karting career when he took home the esteemed Danish Junior Karting Championship. This triumph signaled a turning point and confirmed his reputation as a future motorsport star. His success in karting opened doors for him to compete in more demanding events, including the European Karting Championship, against some of the top young drivers worldwide.

Intense rivalry, education, and development defined these crucial years in karting. From mechanical problems

Kevin Magnussen Biography

to the strain of high-stakes races, Kevin encountered several obstacles that shaped his driving style. As he regularly raised his performance, his capacity for adaptation and fast learning became clear.

Magnussen's karting adventures taught him the virtues of discipline, tenacity, and fierce competitiveness. Driven by the great need to succeed and the awareness that his aspirations were within reach, he was well-prepared to meet the challenges ahead by the time he switched to motor racing. Kevin Magnussen's early racing experience set the stage for his future Formula 1 triumph, hence defining him as a fierce competitor and a real Viking Warrior.

CHAPTER 2: CLIMBING THE MOTORSPORT LADDER: FROM FORMULA FORD TO FORMULA RENAULT

Following his triumphant karting career, Kevin Magnussen's climb to the motorsport ladder started in real force when he moved to car racing in the early 2010s. His first significant action was joining the Danish Formula Ford Championship in 2010. A vital training ground for young drivers hoping to establish themselves in motorsport, this series was renowned for its economy.

Kevin showed extraordinary ability and tenacity in his first Formula Ford season, rapidly rising to be a strong rival. He regularly placed on the podium and won several races, ranking second in the championship. Team

Kevin Magnussen Biography

managers and sponsors were drawn to his outstanding performances, which prepared him for his next racing step.

Magnussen advanced to the British Formula Ford Championship in 2011 to continue displaying his abilities. Having to contend with a bigger and more varied field, he quickly picked up the difficulties the series posed. His diligence and dedication paid off when he grabbed the championship title, confirming his status as one of the most brilliant young drivers in motorsport.

Kevin attracted the attention of elite teams and sponsors after winning his title in Formula Ford, enabling him to advance in his career. He made the big move to the Formula Renault 2.0 class in 2012, which is absolutely vital for drivers hoping to get into Formula 1. Many in the motorsport scene noticed him this way. Formula Renault was recognized for turning out multiple top-notch Formula One drivers.

Kevin Magnussen Biography

Magnussen competed in the Eurocup Formula Renault 2.0, thereby facing more competition. Rising to the test, he displayed his speed, talent, and racecraft. 2013 marked his best season when he dominated the competition, accumulating several wins and finally winning the Eurocup championship. This success not only strengthened his reputation as a future star but also drew the interest of Formula 1 teams ready to find fresh potential.

Kevin's family's motorsport experience shaped his approach, which was marked during these early years in car racing by a strong work ethic and a ferocious competitive spirit. His exploits in the Formula Ford and Formula Renault championships gave him vital racing strategy, car setup, and tire wear management skills knowledge that would prove priceless as he entered the world of Formula 1.

Magnussen's path through the ranks of motorsports is a shining example of the commitment he will need to thrive in a very tough field. His achievements in Formula

Kevin Magnussen Biography

Ford and Formula Renault prepared him for his ultimate Formula 1 debut, therefore launching his journey to become a household figure in the sport.

Early Victories And Building A Reputation

Kevin Magnussen's early successes were very important in building his reputation as a great racing talent as he moved up the motorsport levels. Following his domination of the Danish Formula Ford Championship, he established himself in the British Formula Ford Championship in 2011 with a series of remarkable performances, including several race victories. Both team owners and rivals respected him for his skill and versatility, as he could regularly perform under duress.

Magnussen's career reached a major turning point in 2012 with his effective switch to the Eurocup Formula Renault 2.0 championship. Against some of the top young drivers in Europe, he proved his ability right

Kevin Magnussen Biography

away. Although he finished fourth overall in his rookie season, he really came shining in 2013. Claiming an amazing six victories and many podiums over the season, he claimed a blend of aggressiveness and accuracy, therefore winning the championship. This success not only showcased his ability but also established him as among the top candidates for the next Formula 1 team.

Major motorsport teams and sponsors noticed his exploits in the Eurocup Formula Renault series, therefore confirming his status as a driver to see. His accomplishment attested to his work ethic, strategic thinking, and capacity to maximize his car's performance. Beyond merely winning races, Magnussen's successes demonstrated his ability to control tire wear, grasp race dynamics, and execute calculated overtakes all vital abilities in the high-stress environment of professional racing.

Kevin grew close to fans and the media as he continued to accumulate successes. Many were drawn to him because of his kind attitude and love of the game, which enhanced his profile. By the time he qualified for Formula 1, Magnussen was not simply another rookie; he was a well-liked racer with an evolving legacy.

These early successes and the reputation he developed shaped his career path. They laid the groundwork for his move to Formula 1, where he would encounter even more difficulties and keep proving the Viking Warrior attitude that defined his racing career.

CHAPTER 3: SIGNING WITH YOUNG DRIVER PROGRAM OF MCLAREN

With Kevin Magnussen signed under McLaren's Young Driver Program, his developing racing career underwent a turning point, and he had a road to the top of motorsports. Following a great run in the Eurocup Formula Renault 2.0 series, where he won the title in 2013, several Formula 1 teams noticed his outstanding performance. But it was McLaren, a venerable organization with a great track record of nurturing young talent, that saw his promise.

Magnussen signed formally for McLaren's Young Driver Program in late 2013, joining ranks with other gifted drivers, including Lewis Hamilton and Jenson Button, both of whom had previously benefited from the program. For Magnussen's career, as well as for

Kevin Magnussen Biography

McLaren, which sought to raise the next generation of racing stars, this alliance was important. The program gave him access to technical support, advanced training, and chances to test Formula One cars vital tools.

During his time with McLaren's Young Driver Program, Magnussen participated in numerous tests and simulator sessions, where he acquired priceless knowledge by collaborating closely with engineers and veteran drivers. His results on these tests confirmed McLaren's confidence in his ability, which resulted in his ultimate elevation to a race seat for the 2014 Formula 1 season.

For Magnussen, a team with a history of success in Formula 1, the chance to drive for McLaren was a dream come true. Along with providing a stage to showcase his abilities, his affiliation with the team put him under the direction of seasoned experts who could assist him in negotiating the complexity of Formula One racing.

Excitement about Magnussen's debut surrounding the 2014 season as many people wanted to see how the young Danish driver would do internationally. Signing with McLaren's Young Driver Program was more than simply a stepping stone; it was evidence of his diligence, will, and awareness of his future star potential in Formula 1. This turning point prepared him for his arrival into the sport, where he would immediately have an impact and carry on his path as a Viking Warrior in motorsport.

Pressures And Preparation: The Road To Formula 1

The stakes were great as Kevin Magnussen got ready for his path to Formula 1. Following a place in McLaren's Young Driver Program, he was under constant pressure and expectation. The journey from minor formats to the height of motorsport is always difficult. For Magnussen, it required assuming the role of eminent forebears like Lewis Hamilton and Jenson Button.

Kevin Magnussen Biography

As he became engrossed in the technical side of F1, training got more intense. Spending several hours in the simulator, he honed his skills in readiness for the challenging circuits ahead. He learned the nuances of McLaren's car. The weight of family legacy added to the pressure, not just from the team and supporters. Given a father who had competed at the top levels of motorsports, Kevin was expected to shine.

Magnussen stayed concentrated and motivated despite the degree of preparation required. To learn to control the psychological elements of racing, he consulted his father and other seasoned racers. Dealing with media attention and the great expectations that accompany being a rookie in Formula 1 was among these. Every action he took would be under a microscope, hence his resolve to maximize his chances.

The excitement developed as the 2014 season got ready. Magnussen debuted formally during the winter testing sessions, where he picked up the F1 vehicles' speed and intricacy right away. His performances showed both his aptitude and fast learning abilities. But the strain only got stronger as he got ready for his first Australian race.

Magnussen plunged into a flurry of emotions as the lights went out at the Australian Grand Prix. Though it was thrilling, racing at the highest level carried the weight of expectation as well. He gave a fantastic performance in the first race of the season, ranking third on the podium, a first-year rookie's outstanding feat. Along with quieting some of the uncertainty, this first triumph confirmed his competitiveness in the sport.

Magnussen's road to Formula 1 proved evidence of his tenacity, will, and pressure resistance. Every difficulty he encountered only sharpened his will, and his early sports experiences set the stage for a lifetime of vigorous pursuit of perfection. Lessons gained during this vital preparation phase would help him greatly negotiate the complexity of the racing scene as he kept competing in F1.

Kevin Magnussen Biography

Debuting In Style: The Stunning Australian Grand Prix Podium

Kevin Magnussen made an absolutely amazing début at the 2014 Australian Grand Prix, therefore marking a turning point in his racing career. The first race of the Formula 1 season was much awaited, and all eyes were on the young Danish driver as he turned on the grid for his debut run with McLaren.

Magnussen showed amazing grace and control from the time the lights went out. Showcasing his aggressive driving style and ability to adapt to the high-stress atmosphere, he deftly negotiated the small pack of automobiles. Notwithstanding the difficulties of a rookie's first race, he stayed focused and ran his race exactly.

Magnussen battled fiercely for a position as the race went on, showing great skill and racecraft against more seasoned drivers. His obvious ability to manage the car in many situations and make split-second judgments on the race won him respect from his rivals and compliments from the pundits.

The turning point of his remarkable debut came when a sequence of events happened before him, allowing him to profit from the suffering of others. A sharp eye for chance and quick throttle control helped him to move into a podium spot and cross the finish third. Since his father, Jan Magnussen, stood on an F1 podium in the 1998 season, he is the first Danish driver to do so.

The success was a moment of pride for Denmark and the McLaren squad as much as a personal victory for Magnussen. Being one of the bright prospects in Formula 1, the podium ceremony saw him get the praises of the media and the audience, therefore putting him in front of attention. His outstanding start highlighted his ability and raised expectations for the next events.

Magnussen's remarkable podium at the 2014 Australian Grand Prix was evidence of his diligence, readiness, and relentless support of his family and team. It launched his path as a competitive force in Formula 1. It established him as the Viking Warrior prepared to meet the demands of the sport. Apart from confirming his position in F1 history, even

It provided a launching pad for his career since he kept aiming for perfection in the tough field of motorsports.

CHAPTER 4: THE ROLLERCOASTER ROOKIE SEASON

Kevin Magnussen's 2014 Formula 1 rookie season was an exciting roller coaster with plenty of sensational highs and difficult lows. Following his breathtaking performance in the Australian Grand Prix, when he was placed on the podium, the young Danish driver had great hopes. His performance not only delighted both supporters and experts but also helped to establish him as a future star on the McLaren squad.

Magnussen kept exhibiting his abilities in the next races. Often finishing in the points and establishing himself to be a consistent benefit for the squad, he showed a great capacity to adjust to various circuits. Respect among seasoned rivals came from his clever racecraft and aggressive driving approach; his confidence improved with every race. But as the season went on, Formula 1 started to really kick in.

The mid-season presented difficulties testing Magnussen's fortitude. Technical problems beset the McLaren car, producing varying performance. There were races where he battled to find the speed and finished outside the points. Media and supporters questioned Magnussen's ability to keep the momentum he had developed at the beginning of the season. Both individually and for McLaren, which was in a transitional era with new technical rules, the strain was building, and the stakes were great.

Magnussen stayed driven to show his value in spite of these obstacles. Reminded of the diligence and effort that had gotten him to this degree, he leaned on the experience of his father and staff for encouragement. At the Belgian Grand Prix, when he gave a great effort and once more finished in the points, his perseverance paid off. This outcome restored his confidence and confirmed his competitiveness on the field.

Kevin Magnussen Biography

Magnussen had another challenge when the season came to an end when McLaren declared a driver lineup change for the next year. As he battled to establish his position in the sport, this news added still another level of strain. Though his future was unknown, he entered the last races determined. He completed the season with a solid performance, highlighting his development and adaptability.

The first season Magnussen spent in Formula 1 was an actual rollercoaster. Along the way, this trip had thrilling highs, difficult lows, and priceless lessons discovered. By the end of the season, he had developed himself as a good driver with a bright future ahead. His 2014 experiences helped him to develop personally as well as a racer, preparing him for the difficulties he would encounter in the next few years.

Kevin Magnussen Biography

The Opportunity That Renault In 2016

Kevin Magnussen signed with the Renault Sport Formula One Team, presenting him with a major opportunity in 2016. After a demanding rookie season in 2014 and a second year absent from a race seat, this action signaled a fresh start for the Danish driver and an opportunity to revive his Formula 1 career.

Having bought the Lotus team, Renault had returned to the grid as a full constructor and was keen to rebuild and prove themselves as rivals. Not only did Magnussen provide racing ability, but he also brought experience from his past seasons with McLaren and reserve driving for the team in 2015. The squad found him appealing because of his knowledge of the sport and fast adaptation to new surroundings.

Magnussen embraced the chance with Renault from the start of the 2016 season, eager to help the team grow.

Kevin Magnussen Biography

Though the car was not yet top of the grid, he entered every race determined to maximize his seat. He regularly displayed his ability and racing sense over the season, usually finishing in the points despite the car's performance restrictions.

Magnussen's high point of the 2016 season came at the Belgian Grand Prix, where he finished shockingly sixth. This outcome emphasized his skill in negotiating the complexities of racing strategy and tire management and his capacity to maximize the car's capabilities. The team and the motorsports community praised his performances, therefore confirming his reputation as a driver able to perform under duress.

The alliance with Renault also gave Magnussen a more encouraging environment in which he could work closely with engineers to create the car. He was very important for the team's attempts to improve performance since he provided insightful comments and analysis that would help guide the future of the vehicle.

It became evident as the season went on Magnussen had discovered a new home with the Renault team. His diligence and fortitude paid off; he was able to get a contract extension for the next season. Along with revitalizing his career, this chance with Renault prepared him for more success in the cutthroat Formula 1 scene. It was a turning point that proved he could overcome obstacles and strengthened his will to reach his sport-related targets.

Battling To Prove Himself Once Again

After signing for Renault in 2016, Kevin Magnussen was resolved to show himself once more as a fierce Formula 1 challenger. After a turbulent couple of seasons, he had to prove his value to the squad while also proving his driving ability on a worldwide scene. This new chapter was about atonement, and he went into it totally committed to success.

Kevin Magnussen Biography

Early on in the 2016 season, Magnussen rapidly became enmeshed in fierce competition, frequently facing veteran drivers and upcoming stars. Though not now a front-runner, the Renault car had promise, and he was keen to maximize every bit of performance from it. In wheel-to-wheel racing, Magnussen showed amazing ability, highlighting his aggressive driving approach and will to occupy top spots on the track.

Various obstacles along the season tried his will. There were races when technical problems dogged the automobile, restricting his capacity for performance. Magnussen, though, used these events as teaching and adaptation tools rather than as demoralizing events. His capacity to remain concentrated under pressure revealed his developing driving maturity.

Magnussen regularly found himself in close quarters fighting with other rivals as the races progressed. He was not hesitant to take big actions; he frequently participated in exciting duels, displaying excellent racecraft. His performance in these events confirmed his position in the sport by winning respect from other drivers as well as spectators. His sixth-place finish at the Belgian Grand Prix was particularly remarkable. It demonstrated his ability to exploit possibilities even in cases of less-than-ideal car performance.

Magnussen's will to prove himself went beyond race performance to include securing his place within the Renault team. Working closely with engineers, he gave the team comprehensive comments that helped the car improve throughout the season. His understanding of race strategy and tire control helped the team grow and promoted a cooperative environment.

Magnussen's tenacity paid off as the 2016 season developed. The team noted his constant performance and grit, and his contract was extended for 2017. This confirmation of his ability was evidence of his diligence and a definite sign that he had battled to show himself once again effectively.

Kevin Magnussen Biography

Kevin Magnussen not only found his footing in Formula 1 but also became a major actor in Renault's plans by the end of the season. His path confirmed his persistence and love of racing, therefore setting a strong basis for the next stage of his career and supporting the Viking Warrior attitude that defined his approach to the sport.

CHAPTER 5 : SIGNING WITH HAAS F1 TEAM IN 2017

Signing with the Haas F1 Team in 2017, Kevin Magnussen made a big turn in his Formula 1 career. For the Danish driver, who was keen to expand on his experiences at Renault and establish himself as a major rival in the sport, this action signaled a fresh chapter. Having joined the grid as an American team in 2016, Haas was looking to improve its driver lineup and chose Magnussen because he combined skill and experience.

Joining Haas filled me with enthusiasm and hope. Finishing mid-table, the club had made a strong impression in its first season. It was obviously driven to keep pushing for points. Magnussen saw this as a chance not only to highlight his abilities but also to help a Formula 1 team become established and grow.

Kevin Magnussen Biography

Magnussen sought to leave a big impression right from the start of the 2017 campaign. Quickly adjusting to the Haas car, which Ferrari designed with assistance, he profited from a competitive chassis and a strong engine. His pre-season performance was encouraging, which prepared him for what would turn out to be a defining year for the squad and him.

Magnussen often performed well as the season progressed, often challenging seasoned drivers and enabling Haas to accumulate significant Constructors' Championship points. As he negotiated the competitive field, his aggressive driving style and willingness to take chances were clear-cut indicators of his ability as a strong competitor. He finished in the points several times, notably with a strong performance at the Austrian Grand Prix, where he crossed the line in sixth position.

The collaboration with Haas also gave Magnussen conducive surroundings where he could flourish. The squad promoted a cooperative environment, which let him engage directly with engineers to improve the car's performance and circuit adaptation. As they worked to make the car better all season, this synergy was vital, and Magnussen's advice was much sought.

A fresh will and drive defined Magnussen's time with Haas. Within the squad, he felt a great sense of belonging; his performances mirrored this. He had gained respect among his colleagues by effectively carving out a position for himself as a dependable driver skilled at obtaining the maximum potential from the automobile.

By the end of the 2017 campaign, Magnussen had not only confirmed his place at Haas but also set the foundation for upcoming success. His joining the team marked a turning point in his career since it allowed him to show his skills consistently and show that he was ready to face Formula 1 once more. The alliance signaled the start of a new era in which the always-changing terrain of motorsports would present chances for development and competition.

Kevin Magnussen Biography

The Highs And Lows Of Driving For A Midfield Teams

Driving for a middle-ground squad such as a Haas F1 squad offers a different set of possibilities and difficulties than driving the front of the grid or rear. For Kevin Magnussen, the experience was a rollercoaster of highs and lows that molded his development as a driver and tried his will in the very competitive world of Formula 1.

Driving for a midfield squad usually brings highs from the excitement of competition and the chance to participate in close-quarters skirmishes with other drivers. On the track, Magnussen had several thrilling events; he usually found himself right in the middle of the action. Often involving fierce wheel-to-wheel combat with opponents, races demanded every overtaking move to be vital. These fights gave him an opportunity to highlight his talents and racecraft in addition to being thrilling. Finishing in the points was a

Kevin Magnussen Biography

hailed success since it honored both team strategy efficacy and personal ability.

Being on a midfield squad also let Magnussen be instrumental in the car's development. Unlike front-running teams that might concentrate just on maximizing performance for podium finishes, midfield teams usually depend on driver comments to make major changes. Working closely with engineers to fine-tune the car's setup, Magnussen's thoughts were priceless as he enabled the team to adjust to different circuits and increase general performance. As he felt personally engaged in the team's path toward success, this cooperative effort developed camaraderie and purpose.

But the lows of driving for a middle-ground squad were equally strong. The ongoing fight for points meant that outcomes might be erratic. Magnussen and his crew completed their plans well in certain races, only to have dependability problems, bad weather, or errors from other teams ruin their efforts. Such failures were aggravating since they sometimes meant that preparation and diligence went unpacked. A driver may find great emotional weight in changing fortunes, especially in a highly competitive market with narrow margins.

Furthermore, even if the midfield could present exciting racing and development chances, it usually needs more means and funding from the top teams. This difference can make it challenging for a driver to guarantee constant performance since the car's performance cannot always coincide with the front-runners'. Magnussen faced difficulties racing in a car that, although competitive, needed more constant ability to contend for top places.

The great scrutiny drivers endured added still another level of complication. Being on a midfield squad meant that errors could be recognized and attacked in a sport where every action is examined. Magnussen knew that any error might eclipse the points he had secured, underlining the need to perform well constantly.

Kevin Magnussen Biography

Magnussen skillfully negotiated these highs and lows with fortitude over his stay at Haas. Every event from a great points finish to a disappointing race helped him develop as a driver. Driving for a midfield team tested not just his abilities on the track but also his mental toughness, therefore preparing him for the erratic nature of Formula 1. In the end, the road with Haas helped him become a respected rival in the sport, able to flourish in the face of difficulty and grab possibilities as they presented themselves.

CHAPTER 6: ICONIC BATTLES AND RIVALRIES

Over his Formula 1 career, especially with the Haas F1 Team, Kevin Magnussen became known as a fearless, forceful driver. His unrepentant attitude to racing, marked by audacious overtaking moves and a dogged search for positions on the track, shaped this image.

Magnussen has shown a natural inclination to take chances from the start of his career. In his debut season, where he frequently engaged in exciting wheel-to-wheel battles with seasoned drivers, his aggressive racing style was clear. This reputation only strengthened once he moved to Haas, where he regularly found himself caught in fierce duels with rivals, frequently producing thrilling on-track events that delighted both spectators and analysts.

Kevin Magnussen Biography

Tight racing situations especially revealed Magnussen's bravery. Often plunging down the inside or making bold movements on the outside, he showed an amazing ability to position his car in areas many would regard as too dangerous. This natural ability for strong overtaking not only stunned his competitors but also demonstrated his self-confidence and will to stand out in the marathon. His aggressive moves were a trademark of his racing character, which won him a reputation as a racer ready to challenge limits.

Many unforgettable fights that emphasized his unwavering attitude helped to strengthen his name. Magnussen participated in hair-raising fights in many races, showing extraordinary car control and racecraft. One particularly noteworthy instance was during the 2018 Bahrain Grand Prix, when he battled bravely to hold his place against experienced drivers, therefore displaying his toughness and refusal to compromise. Such actions helped him to establish himself as a driver who would not back down, independent of the consequences.

Kevin Magnussen Biography

This strong approach came with some issues. Magnussen's racing was sometimes criticized for being overly strong. There were times on track when his aggressive style resulted in crashes or controversial events that spurred discussions on his strategies. Still, he stayed unreserved about his method, sometimes contending that racing is essentially about taking chances and that his approach was essential for his competitive temperament.

Magnussen's fearlessness also transcended the course, as he often expressed his opinions about racing and his fellow rivals. His direct approach and eagerness to express his ideas made him a divisive person, adding to his reputation as a driver not hesitant to question the existing status quo. Fans who valued his authenticity and love of the game found resonance in this openness.

Magnussen kept embracing this brave image as his career developed. One of the most thrilling drivers to see in Formula 1, his forceful approach became a defining quality. Whether fighting for points in a midfield vehicle or against the top in the sport, he stayed dedicated to stretching the envelope of what was feasible on the circuit.

Kevin Magnussen's toughness, strong racing instincts, and relentless devotion to competition ultimately helped him to build a reputation as a bold, forceful driver. This image not only made him popular among supporters but also confirmed his position in the cutthroat world of Formula 1, therefore defining him as a driver to an admirer.

The Art Of Fearless Overtaking: Kevin's Racing Style Analyzed

Kevin Magnussen has participated in several legendary races and clashes over his Formula 1 career that have shown his aggressive racing approach and will on the track. These rivalries have not only highlighted his abilities but also helped to establish his reputation as a brave driver ready for exciting on-track duels.

Kevin Magnussen Biography

Among Magnussen's most prominent rivalries have been those with his fellow midfield players. With each interaction between him and drivers like Sergio Pérez, Lando Norris, and Daniel Ricciardo typified by intense wheel-to-action, his bouts typically played out during pivotal points in events. Close racing, daring overtakes, and clever defense were common features of these contests, which created unforgettable events for viewers and added to the drama of every event.

One very unforgettable fight Magnussen participated in was the 2018 Azerbaijan Grand Prix, where he was in a high-stakes scenario alongside Romain Grosjean, his Haas teammate. As the two drivers battled for position in the wild race, there was obvious conflict between them. Magnussen drove aggressively, making strong moves and furiously defending against Grosjean. This battle among the midfield drivers fighting for the same points highlighted the competitive nature existing not only among colleagues but also inside the squad.

Kevin Magnussen Biography

Magnussen's interactions with Pérez were likewise noteworthy, particularly in the 2019 season. Often caught in fierce contests for vital points, the two drivers had racing marked by tight calls and aggressive on-track interactions. These meetings reflected the intense rivalry in the midfield, as every position acquired may have a big impact on the championship results.

Lando Norris became another major adversary for us, particularly in the 2020 season. Their struggles sometimes took place in the closing phases of races, when both drivers strained their boundaries in quest of points. Fans saw a mix of techniques and plans competing head-to-head. Magnussen's aggressive driving collided with Norris's cool and deliberate approach, creating fascinating viewing.

Apart from this rivalry, Magnussen's interactions with seasoned talents like Lewis Hamilton and Sebastian Velle also stand out. These fights gave a peek at his potential against the top of the sport, even if their frequency was lower because of different car

Kevin Magnussen Biography

performances. They often led to exciting events when Magnussen found himself accompanying these champions and proving his own against the greatest: Magnussen's bold approach.

Apart from defining Magnussen's career, his rivalries and conflicts have added to the thrill and uncertainty of Formula 1 racing. Every interaction with rivals or teammates has added to his legacy as a driver who is not hesitant to fight for every position and is ready to participate in the fierce competition defining the sport. These legendary events have confirmed his reputation as a tough midfielder, guaranteeing that supporters will always remember him as a driver who gave every race emotion and intensity.

CHAPTER 7: THE 2020 SEASON AND THE CHALLENGES OF RACING AMIDST A PANDEMIC

Unprecedented obstacles brought about by the COVID-19 epidemic defined the 2020 Formula 1 season from all angles. Navigating this particular terrain for Kevin Magnussen and the Haas F1 Team needed flexibility and resilience both on and off the course.

Lockdowns and restrictions placed on clubs worldwide delayed the season's start. Conventional pre-season testing was greatly truncated, leaving drivers without enough time to prepare. For Magnussen, this was especially difficult since the limited running time meant

Kevin Magnussen Biography

fewer chances to adjust the car and become used to any off-season adjustments.

When racing at last started, it did so under rigorous health guidelines. Teams used strict testing and social separation policies while races were staged without spectators. These surroundings changed the nature of the sport even while it guaranteed the safety of everybody engaged. For drivers and teams both, the lack of roaring crowds and the customary buzz around the events produced an unusual experience. Magnussen would sometimes consider how unusual it felt to compete in deserted tracks, a far cry from the electrified environment usually defining Formula 1.

On the track, Magnussen had major difficulties over the season. Haas battled with car performance and dependability, so securing points proved hard even as he displayed an aggressive racing style. Magnussen had previously been a regular point scorer for the squad, so this was especially irritating. The season's challenge was exacerbated by a car's underperformance combined with the extreme pressure of racing in such a different environment.

Magnussen stayed attentive to these challenges to maximize the circumstances. Drawing on his knowledge, he gave the engineering team insightful criticism that would help them negotiate the car's restrictions. Particularly when the epidemic drove teams to be more creative and imaginative in their approaches to racing strategy and development, his capacity to change with the times was vital.

The difficulties of the 2020 season culminated in the choice Magnussen would make that would affect his career going ahead. As the season came to an end, Haas said they would not keep him for the next year, choosing a fresh driver lineup. For Magnussen, this choice was mixed since it signified the end of his tenure with a team that had grown to be important for his Formula 1 path.

For Kevin Magnussen, the 2020 season tested his moral fiber. Running during a pandemic posed special difficulties that tested his will and fortitude. Although it was a challenging year in many ways, it also proved evidence of his commitment to the sport and his relentless love of racing, therefore preparing the ground for the next stage of his career in motorsport.

Departure From Formula 1

Kevin Magnussen's resignation from Formula 1 at the end of the 2020 season signaled a major career turning point. After six seasons in the sport, he saw the Haas F1 Team split ways with mixed feelings that reflected the difficulties he encountered during his career.

Magnussen had become known as a gifted and forceful driver over his Formula 1 career. From his great rookie podium finish at the 2014 Australian Grand Prix to the challenges of the 2020 season, where the performance of the Haas vehicle hindered his capacity to compete for points regularly, he went through highs and lows. It was abundantly evident as the season came to an end that the

Kevin Magnussen Biography

squad was trying to update its roster in order to develop going forward with fresh talent.

Many supporters and onlookers were taken aback by Magnussen's news of his leaving since, in his early years, he had displayed flashes of genius. Still, it also mirrored the reality of Formula 1, where driver contracts are frequently set by team chemistry, performance, and commercial considerations. For Magnussen, this meant leaving a team he had developed a close relationship with but finding irregular success in the later years.

Magnussen felt conflicting emotions following the announcement. He was happy with his chances in Formula 1 and with the time he had spent with the team and other drivers. Still, he felt let down at not being able to display his best in a competitive car. As he considered his next professional move, he was much troubled by the unknown future.

Though leaving Formula 1 disappointed, Magnussen stayed steadfast in his love of motorsports. The excitement of racing has always motivated him, so the hurdles he encountered did not affect his competitive nature. He investigated several motorsport prospects, and conversations regarding his future kept his options open for possible activities outside of Formula 1.

Kevin Magnussen's departure from Formula 1 finally signaled the end of an era, but it also brought fresh opportunities. Although he might have turned away from the height of motorsports, his reputation as a bold and forceful driver will always be a major component of who he is. His path in Formula 1 set the groundwork for whatever lay ahead, assuring that his love of racing would always propel him onward in the next phase of his life.

CHAPTER 8: THE UNEXPECTED COMEBACK WITH HAAS IN 2022

Fans and the motorsports community noticed Kevin Magnussen's surprising return with the Haas F1 Team in 2022, which marked a drastic change of events. Following a year away from Formula 1 and engaging in other racing activities like sports car racing, Magnussen answered a call that would restart his F1 career.

The chance presented itself when Haas chose to split from Nikita Mazepin not quite before the season started. The geopolitical environment following Russia's invasion of Ukraine, which resulted in the sponsorship agreement with Mazepin's father being terminated, had a major influence on the choice. Haas went to Magnussen, a driver who had before raced for the team and had already developed a close relationship with the crew, when the seat became unexpectedly free.

Recognizing the offer as a second chance in a sport he had always loved, Magnussen was eager to take it. Not only from Haas but also from supporters who valued his strong driving approach and the combative attitude he brought to the grid, the news of his comeback excited everyone. The story of a driver going back to a squad he had already competed in gave the next season more intrigue.

Magnussen had to reacclimate to the demands of Formula 1 as the 2022 season started. He had to adjust to the new rules, altering vehicle performance traits and design. Still, his knowledge of the team helped him pick things right away. His performance during pre-season testing surprised the squad and onlookers, suggesting that he was prepared to once more compete at a top level.

Kevin Magnussen Biography

Starting brilliantly for Magnussen in the 2022 Bahrain Grand Prix, when he came in an amazing fifth place, the season not only marked his comeback to the sport but also a great outcome for Haas. His performance defied expectations and indicated that he was back in business since it highlighted his inherent skill and capacity to get the best performance from the vehicle. Fans celebrating his comeback and the squad enjoying the increased competitiveness clearly exuded delight about this return.

Magnussen kept surprising throughout the season, regularly scoring points and helping Haas to recover in the Constructors' Championship. Often in the heart of the action, his aggressive driving technique was evident as he battled fiercely with competitors. His background proved his capacity for adaptation and performance under duress as he negotiated difficult race conditions.

Still, the return presented certain difficulties. Formula 1's competitive environment meant that races might be erratic and that occasionally, team plans and car dependability did not work as expected. Still, Magnussen stayed tenacious and focused, exhibiting the will that had shaped his career.

Magnussen's comeback to Haas in 2022 not only restored his career but also gave the team fresh hope. Embracing the chance to once more compete at the greatest level, the cooperation displayed his toughness and love of racing. His comeback narrative became evidence of the erratic nature of motorsports, showing that even a brief absence from the grid might result in a triumphant return with will and ability.

New Challenges And Proving Himself Once Again

Kevin Magnussen re-entered Formula 1 with Haas in 2022. He encountered a set of fresh problems that tried his will, flexibility, and competence. Having missed a year from the sport, he had to prove he still had the competitive edge and will that defined his early Formula 1 years.

Kevin Magnussen Biography

Relearning the particular needs of F1 racing proved to be one of the most immediate obstacles. With modifications to aerodynamics and tire criteria, the new rules unveiled in 2022 fundamentally changed car designs. This forced Magnussen to learn the subtleties of the revised Haas car and quickly adjust to a different driving style. Fortunately, his acquaintance with the team and the engineering crew enabled him to negotiate this change more naturally than many would have predicted.

Magnussen also had to negotiate higher expectations. Not only did his amazing comeback performance at the Bahrain Grand Prix where he placed fifth raise a high standard for the team as well as for him, but for Haas, this first victory inspired fresh optimism and ambition, and the pressure to keep that momentum clear-cut. Every race turned into a test of his ability to produce consistent performances; Magnussen knew he had to prove to critics he still belonged at the highest level of motorsports, so securing his position on the team.

Kevin Magnussen Biography

Magnussen ran against many difficulties on track over the season. The intense rivalry in the midfield meant he was frequently caught in close-quarters fighting with other drivers, notably familiar adversaries Lando Norris and Sergio Pérez. Every race provided fresh challenges and chances; hence, he had to remain sharp and concentrated. Often finding inventive ways to make up positions and defend against opponents, he showed his usual aggressiveness in overtaking moves. His reputation as a brave competitor was built on this dedication to sprinting hard, which also helped him to win respect from his colleagues.

Formula 1's unpredictability also provided another level of difficulty. Events, dependability problems, and competing teams' different performances could significantly change the race result. When the car did not run as predicted or when racing plans did not work out for Magnussen, he became frustrated. Still, he was tenacious, always pushing the crew and himself to run the car to its utmost.

Kevin Magnussen Biography

He was reestablishing himself in a sport that had changed. At the same time, his absence presented Magnussen with difficulty off the track. Along with keeping a good mood and proving his dedication to the sport, he had to reestablish ties to the media, supporters, and the larger F1 community. Fans came to like him because of his openness and genuineness; many valued his path of development and tenacity.

Magnussen not only proved himself again but also confirmed his reputation as a valued Formula 1 driver through diligence and determination. By the end of the 2022 season, he had shown that he was not only a returning driver but also a strong competitor fit for the fast-paced motorsports scene. Fans connected with his path of conquering fresh hurdles and reaffirming his place in Formula 1, therefore confirming his history as a determined driver dedicated to succeeding on the track.

Kevin Magnussen Biography

Adapting To A Changing Sport And Evolving As A Driver

Particularly with his surprising comeback to Formula 1 in 2022, Kevin Magnussen's career has been defined by his ability to adapt to the quickly shifting terrain of Formula 1. Shifting rules, technology developments, and a more competitive field defined F1's progress; hence, he had to not only welcome change but also grow as a driver to remain relevant and successful.

Magnussen immediately had to acclimate to the new technical rules implemented in 2022 when returning to Haas. These modifications changed tire designs, automobile aerodynamics, and general performance qualities, so he had to modify his driving technique quickly. Understanding he needed to change his strategy, Magnussen concentrated on learning the nuances of the revised Haas car and closely coordinated the engineering team to identify the best configuration. His dedication to learning and progress is shown in his eagerness to interact with engineers and offer insightful criticism.

The 2022 campaign also underlined the need for flexibility in race strategy. Given the erratic nature of race conditions including changing weather and tire degradation Magnussen often had to act quickly. Accepting this challenge, he showed a great capacity to adapt his driving strategy on the course in reaction to evolving conditions. His experience enabled him to stay cool under duress. This vital ability helped him negotiate close calls and seize possibilities as they presented themselves.

Magnussen came to see the value of mental fortitude as he relaxed into the cadence of racing. Competing in Formula 1 can have strong emotional highs and lows, and returning after a year away came with particular demands. To adjust, he concentrated on keeping a good attitude and creating plans to handle the unavoidable difficulties of competing at the top level. This mental strength was crucial since it helped him to remain driven

Kevin Magnussen Biography

to reach his objectives and manage disappointments graciously.

Magnussen's experiences with younger drivers, especially those who had started the sport while he was away, added yet another level of adaptability. He took lessons from their fresh viewpoints and used his own experience to guide and offer ideas. This information sharing helped him adopt fresh ideas while honing his methods and promoted friendship among the participants.

Furthermore, the changing character of fan participation in the sport forced Magnussen to adjust off the course. Drivers now interact with their audience on digital channels and social media differently. Accepting this transition, Magnussen started posting his experiences, ideas, and behind-the-scenes peeks into his life as a driver on these sites more actively. This openness not only won him over supporters but also strengthened his brand in a cutthroat market.

Kevin Magnussen Biography

Kevin Magnussen has shown remarkable capacity to develop as a driver and compete in the shifting Formula 1 scene through these several modifications. His path shows a dedication to learning, perseverance, and adaptability all of which are essential skills for success in a sport that always challenges performance and technological capability. Magnussen's development shows not just his development but also the dynamic character of Formula 1 itself as he keeps negotiating the obstacles of racing.

CHAPTER 9: KEVIN MAGNUSSEN'S LIFE OFF THE TRACK

Off the course, Kevin Magnussen portrays a complex guy who strikes a balance between his high-stress Formula 1 career and personal hobbies, family life, and a passion for motorsports outside of racing. Although the demands of racing, travel, and rigorous preparation define most of a Formula 1 driver's life, Magnussen has developed a happy existence apart from the sport.

His family is the center of his life off the course. Kevin is wed to Louise, his long-time friend, and their tight relationship has given him help all through his racing career. Their bond is based on a common love of motorsports since Louise has frequently traveled with him to races and helped him in many activities. Their daughter together delighted Magnussen's life and added

Kevin Magnussen Biography

a fresh perspective. On social media, he regularly posts peeks of their family events to emphasize the value of family in providing stability among the demands of his career.

Beyond Formula 1, Magnussen is fascinated with motorsports. Deeply enthusiastic about all kinds of racing, he regularly takes part in sports car racing and endurance competitions. This participation helps him to keep his connection to racing while investigating several facets of the activity. His social media posts, where he offers insights and experiences from several racing disciplines, thereby engaging with people who share his interest, clearly show his love of motorsports.

Magnussen enjoys training and maintaining a good lifestyle well beyond racing. Formula 1 requires drivers to be in top physical shape, so he takes great care of this. Strength training, cardio, and other workouts catered to improving his performance on the track constitute part of his fitness schedule. Magnussen also stresses the need for mental fitness since he realizes that, in fact, the

Kevin Magnussen Biography

psychological component of racing is just as vital as the physical training.

Apart from his obligations to racing and family, Magnussen is renowned for his interesting demeanor and sense of humor. Often sharing lighthearted events from his life, he has built a reputation for being approachable and grounded. Because of his relatability, he is a popular person on the paddock among the supporters. On social media, he interacts with supporters. He shares his experiences, ideas, and honest events from his life as a driver.

Magnussen also stays close to his Danish background and frequently expresses delight in representing Denmark in the worldwide motorsports scene. He has taken part in activities supporting Danish motorsports and discussed the need to motivate the next generation of racers from his nation.

Kevin Magnussen's life off the course shows a combination of personal pursuits, family values, and a strong dedication to motorsports overall. His path highlights the many facets of a contemporary racing driver and the commitment needed to succeed in Formula 1 while keeping a happy life outside of racing. Driven by his love of racing and interacting with the community that supports him, Magnussen is anchored by his family while he negotiates his profession.

Personal Interest, Family And Life In Denmark

Beyond the racetrack, Kevin Magnussen's interests reflect his diverse character and the several activities he enjoys in his leisure time. Motorsports, in general, are one of his main interests since he often investigates several racing disciplines, from sports cars to karting. This passion helps him interact with many facets of the sport he enjoys. It keeps him in touch with the larger racing community.

Kevin Magnussen Biography

Apart from racing, Magnussen strongly enjoys fitness and outdoor activities. Physical health comes first for him, and he frequently posts bits of his training sessions, including cardio, strength building, and several endurance activities. Maintaining optimal performance in Formula 1 depends on both physical and mental stamina, so his dedication to health is absolutely vital.

Magnussen also enjoys engineering and technology. His passion for how things function fits his background in motorsports since he values the subtleties of car performance and design. Sometimes, this passion shows up in his social media material, where he interacts with supporters on technical racing issues and sports advancements.

Regarding interests, he likes to spend time outside engaging in sports like hiking and cycling, which let him relax and commune with the surroundings. These activities not only give a physical release but also help one reenergize psychologically within the demanding F1 schedule.

Kevin Magnussen Biography

On the personal front, Magnussen is wed to Louise, who has shaped his life and work greatly. Deeply bonded by mutual support and awareness of the difficulties of professional racing, the pair Their bond is strong on the same experiences, usually seen together at races and events, where Louise is always encouraging.

They are parents of a daughter who brings happiness and gives their lives fresh vitality. Magnussen often talks about how parenthood has enhanced his life and given him a viewpoint outside of racing's competitive drive. Despite his hectic schedule, he values family time and emphasizes the need to enjoy special events with his loved ones.

Kevin Magnussen's hobbies and his life with Louise and their daughter show a balance between his career obligations and his position as a husband and parent. His capacity to pursue several interests while raising his family shows that he is a well-rounded person who stays rooted even in the demanding Formula 1 environment.

Kevin Magnussen Biography

The Legacy Of A Racing Warrior

Kevin Magnussen earned the legacy of a racing warrior through a voyage in the realm of motorsports characterized by tenacity, resiliency, and a tireless quest for excellence. From his early days karting in Denmark to his remarkable career in Formula 1, Magnussen has shown a relentless love of racing that appeals to both enthusiasts and aspirant drivers.

His legacy is based on a set of outstanding successes and landmark events that show his on-course ability and perseverance. With an amazing debut at the Australian Grand Prix and a podium finish, a mark that defined his career Magnussen exploded into the F1 world in 2014. Early success not only highlighted his skill but also his ability to compete at the highest level, thereby rapidly establishing him as a strong presence on the paddock.

Magnussen faced many difficulties throughout his Formula 1 career, including the great demands of vying

Kevin Magnussen Biography

for a midfield team and swings in car performance. His reputation as a bold and forceful driver was cemented by his competitive attitude toward overcoming these challenges. Bold overtakes, and a readiness to push the envelope define Magnussen's signature approach, which won him respect among peers and appealed to fans. His story grew to include legendary rivalries and on-track confrontations, strengthening his status as a driver who was not hesitant to fight for any position.

The surprising comeback to Haas in 2022, following a year away from Formula One, demonstrated his tenacity and flexibility. Magnussen's passion and determination were evident in his fast reacquainted with the sport and performance power. His return not only restored his career but also represented the relentless attitude of a racing warrior who defies definition based on adversity.

Beyond his achievements, Magnussen left behind a legacy in character of the course. Well-known for his grounded approach and captivating style, he has developed a close relationship with his supporters through public events and social media. People relate to him because of his transparency and genuineness; he is. Therefore, a relevant person in a sport is sometimes defined by gloss and celebrity.

Beyond the racetrack, Magnussen's impact is felt because he inspires aspirant drivers, especially those from Denmark. His path inspires the next generation to follow their aspirations, regardless of obstacles, by showing the diligence, commitment, and endurance needed to excel in motorsports.

Kevin Magnussen's reputation as a racing warrior will be defined not just by his successes and podiums but also by his perseverance in the face of hardship, his relentless passion for racing, and his capacity to personally connect with fans as he negotiates his career. His narrative is evidence of the fierce rivalry and unwavering quest for perfection, therefore transforming the motorsports industry and motivating the next generations of racers for years to come.

CONCLUSION

Few stories in the great fabric of motorsports are as vivid and dramatic as Kevin Magnussen, the Viking Warrior of Formula 1. His path is evidence of the unwavering spirit of a competitor who has surmounted the most difficult lows and the highest highs while nonetheless following his love of racing. From his early karting days in Denmark to his incredible comeback with the Haas F1 Team, Magnussen has shown not only his obvious ability but also a relentless will to succeed.

Magnussen has faced strong rivals, negotiated the complexity of changing rules, and embraced the demands of top competitiveness across his career. His persistence and forceful driving approach have helped him to establish himself as a courageous racer, one who does not hesitate to meet obstacles and is always ready to grab possibilities. On and off the track, he has constantly shown the heart of a warrior, whether celebrating a great podium result or negotiating the complexities of a midfield struggle.

Kevin Magnussen's legacy resides in his character, transcending the numbers and honors. Aspiring drivers all around find encouragement in him since he epitomizes tenacity. His real love of motorsports, combined with his ability to engage with supporters, has made him a beloved person on the paddock. Through his path, he has demonstrated that resilience, humility, and dedication to one's work define success more so than wins.

It is quite evident from considering Magnussen's career that his story is not over. In an always-shifting sport, the Viking Warrior keeps developing, growing, and welcoming fresh challenges. His love of racing never changes; it drives his desire to challenge limits and motivate the following generation of racers.

Kevin Magnussen's biography ultimately celebrates the pure human spirit rather than only chronicling a driver's

Kevin Magnussen Biography

path through Formula 1. His narrative teaches us that real warriors rise to confront challenges rather than withdraw from them, therefore guiding their way with bravery and will. Kevin Magnussen is a bright example of what it means to be a Viking warrior, a fighter in the arena, a champion of the track, and an inspiration for everyone who dares to dream as he continues to create his history in the realm of motorsports.

www.ingramcontent.com/pod-product-compliance
Lightning Source LLC
Chambersburg PA
CBHW070358230526

45471CB00006B/2632